Generis
PUBLISHING

EFdA (Islatravir): An Outstanding Anti-HIV Modified Nucleoside

-From Conceptualization to Present-Day Clinical Findings-

Hiroshi Ohrui

Title: **EFdA (Islatravir)**: An Outstanding Anti-HIV Modified Nucleoside
-From Conceptualization to Present-Day Clinical Findings-

ISBN: 979-8-89248-176-2

Author: Hiroshi Ohrui

Cover image: www.pixabay.com

Publisher: Generis Publishing
Online orders: www.generis-publishing.com
Contact email: info@generis-publishing.com

EFdA (Islatravir):
An Outstanding Anti-HIV Modified Nucleoside
-From Conceptualization to Present-Day Clinical Findings-

This book is dedicated to the memory of my esteemed mentors, Dr. Masanao Matsui and Dr. Kenji Mori.

Acknowledgements:

The author extends heartfelt gratitude to all colleagues, with special appreciation for Dr. Hiroaki Mitsuya and the team of scientists at Yamasa for their significant contributions to the biological assessment of modified nucleosides.

Contents

Preface:

The rise of numerous viral infectious diseases, such as AIDS, influenza, West Nile virus, SARS, and Covid-19, poses significant challenges to global public health. Consequently, developing effective methods to combat viruses emerges as a crucial area of focus for enhancing human health in the current and future centuries. The need for innovative strategies in this battle is increasingly urgent.

While my expertise lies in organic chemistry, specifically in the realms of nucleoside chemistry, carbohydrate chemistry, and chiral discrimination, rather than virology, I have embarked on a journey to devise a novel approach to combat viruses. My exploration is grounded in the principles of geometrical organic chemistry—a perspective that employs spatial thinking in organic chemical reactions—combined with a foundational understanding of biochemistry and virology, as well as practical experience in nucleoside chemistry.

It's worth noting that I have developed effective methods for chiral discrimination, addressing challenges associated with the diastereomer method (*Hiroshi Ohrui, Analytical Sciences, 24 (1), 31-38, 2008*).

This book aims to convey the strategy I've pursued in the hope of inspiring others. Regardless of one's field of interest, it's possible to achieve impactful results through dedicated inquiry and innovation.

To begin, I wish to share the core principles underlying my work on antiviral modified nucleosides, despite my limited expertise in virology.

The fundamental concept:

Viruses adapt to environmental changes through mutations, which often leads to the emergence of strains that are resistant to current drugs and vaccines. Consequently, mutations have been predominantly viewed as a challenge in antiviral therapy, being associated with the development of drug resistance. However, I posit that mutations also present a significant opportunity for the advancement of antiviral drug development.

Mutations occur when viruses incorporate incorrect nucleosides into their genetic material, revealing that viral nucleic acid polymerases have less stringent substrate selectivity compared to those of humans. Humans do not typically incorporate incorrect nucleosides into their DNA due to the high fidelity of our nucleic acid polymerases.

This difference in substrate selectivity between viral and human polymerases can be leveraged to create modified nucleosides that are selectively integrated into viral genomes. These modified nucleosides could inhibit viral replication without affecting human DNA, thereby offering a targeted and non-toxic treatment option.

Building on this concept, I have formulated several hypotheses aimed at addressing the challenges in treating human immunodeficiency virus (HIV). This approach has led to the successful development of a promising anti-HIV agent, 4'-C-ethynyl-2-fluoro-2'-deoxyadenosine, which is both highly effective against HIV and has low toxicity, as well as prolonged activity. (EFdA). (Fig. 1) [1]

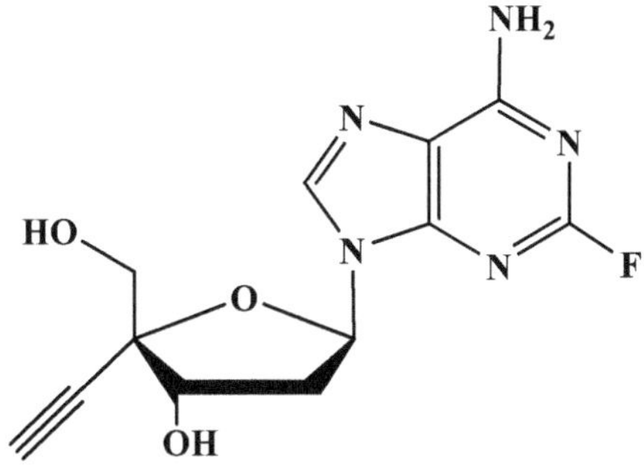

EFdA

Fig. 1

Since 2012, EFdA has undergone clinical trials conducted by Merck & Co., and the company has designated it as Islatravir.

***Albert von Szent-Györgyi's* saying:**

"*Discovery consists of seeing what everybody has seen and thinking what nobody has thought*".

***Louis Pasteur's* saying:**

"*Chance Favors the prepared mind*".

Inspired by various insights, I adopted a unique motto for my laboratory: *"Let's explore and study the unseen that remains invisible to others."*

This motto underscores the significance of interdisciplinary knowledge in uncovering new perspectives and ideas.

My journey into developing anti-HIV modified nucleosides with low toxicity and resistance to HIV mutations began in collaboration with Asahi Breweries Ltd. in 1991. The collaboration was essential for conducting biological evaluations of my modified nucleosides. At that time, Asahi Breweries had recently entered the pharmaceutical research and development sector, coinciding with one of my students joining their team. Our joint research focused on 4'-C-methylnucleosides. However, due to the absence of an anti-HIV evaluation system within the company, Dr. Masanori Baba of Fukushima Medical University, who later moved to Kagoshima University, undertook the evaluation of the anti-HIV activity of these nucleosides.

After Asahi Breweries exited the pharmaceutical industry, I sought the expertise of Yamasa Corporation in Chiba, Japan, for further biological assessment of my nucleosides. Yamasa then enlisted Dr. Hiroaki Mitsuya of Kumamoto University and the NIH to evaluate their anti-HIV efficacy.

In this collaborative effort, while I was responsible for designing and synthesizing the modified nucleosides, the biological evaluation was carried out by Dr. Mitsuya and Yamasa Corporation.

Introduction:

Following Dr. Hiroaki Mitsuya's identification of 3'-C-azido-3'-deoxythymine (AZT) as an effective anti-HIV compound in 1985, a variety of 2',3'-dideoxy nucleoside (ddN) analogues have been synthesized as nucleoside reverse transcriptase inhibitors (NRTIs). Despite their development, HIV strains resistant to these NRTIs have quickly and frequently arisen.

To counteract the issue of HIV resistance, the highly active antiretroviral therapy (HAART) regimen, combining multiple NRTIs with protease inhibitors, was introduced. This strategy has significantly enhanced the life quality and survival rates of HIV-infected individuals.

Nonetheless, HAART faces several critical challenges, including the development of new resistant HIV strains, the necessity for high dosages of medication, and adverse drug reactions. Consequently, there is an urgent need for the creation of new anti-HIV medications that are both highly effective in preventing resistance and have minimal side effects.

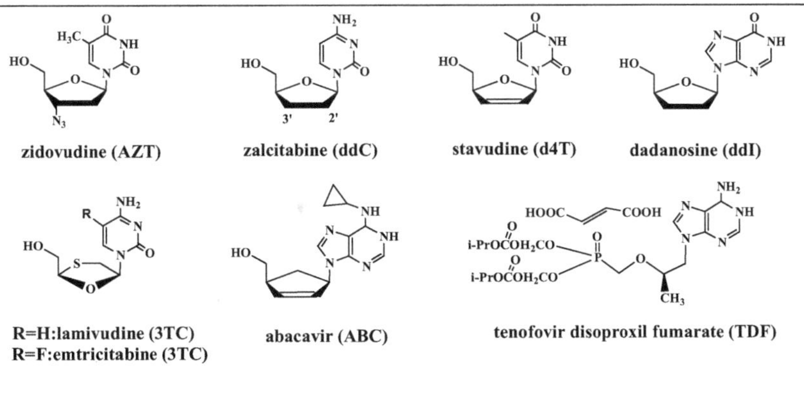

Clinical ddNs

Fig.2

Fundamental Concept for the Development of Anti-viral Modified Nucleosides:

The challenges associated with current HIV treatments led me to contemplate the creation of optimal anti-HIV modified nucleosides capable of thwarting the rise of resistant HIV strains while minimizing side effects. Initially, I derived a basic yet innovative concept for crafting such anti-HIV modified nucleosides, addressing the shortcomings of existing HAART strategies. Surprisingly, this foundational idea, rooted in my basic understanding of virology and biochemistry, seemed overlooked by virologists.

Drawing from the Sanger Method of DNA sequencing, which highlights that 2',3'-dideoxynucleosides (ddNs) act as terminators for DNA polymerases, hence their inherent toxicity, yet their clinical use as NRTIs underlines a differential susceptibility between reverse transcriptase (RT) and DNA polymerases to ddNs. RT is more receptive to incorporating ddNs than DNA polymerases are. This key observation fueled my inspiration for a fundamental concept in developing ideal anti-HIV modified nucleosides. Given its significance, I find it crucial to reiterate this concept here for clarity and emphasis.

The fundamental concept:

Viruses undergo mutations in response to environmental changes, often developing resistance to antiviral medications and vaccines. Consequently, mutations are primarily perceived as a mechanism of resistance within the realm of antiviral therapy. However, I view mutations as a valuable opportunity for crafting antiviral modified nucleosides that are specifically effective against viruses while being harmless to humans, for several reasons.

Mutations occur when viruses incorporate incorrect (unintended) nucleosides into their genetic material, revealing a lack of stringent substrate selectivity by viral nucleic acid polymerases. In contrast, humans do not incorporate incorrect nucleosides into their DNA, indicating a more precise substrate selectivity in human nucleic acid polymerases. This difference in substrate selectivity between viral and human polymerases presents an opportunity to develop modified nucleosides that are preferentially integrated into viral genomes. Consequently, these modified nucleosides could target viruses effectively without harming human cells.

I pondered over the reasons behind the development of HIV mutants resistant to clinically used NRTIs, which are part of the ddN family, and formulated a hypothesis regarding the chemical structure of new NRTIs with high anti-HIV activity capable of preventing drug-resistant mutations. This led to the conceptualization of 4'-C-substituted-2'-deoxynucleosides (4'SdNs) based on this hypothesis.

Additionally, I proposed another hypothesis to reduce the toxicity of nucleoside derivatives, drawing from the initial concept and my previous research on the correlation between the structure of nucleoside derivatives and their biological activity.

The research on the synthesis and biological assessment of 4'SdNs validated these concepts and hypotheses, culminating in the creation of 4'-C-ethynyl-2-fluoro-2'-deoxyadenosine (EFdA). Despite its simplicity, EFdA exhibits exceptionally high efficacy against all strains of HIV, low toxicity, and resistance to enzymatic breakdown, ensuring prolonged activity. Remarkably, EFdA has successfully prevented the development of drug-resistant HIV mutations for approximately 20 years since its introduction.

Working hypotheses to solve the problems in this study:

To address the issues, I have suggested two potential solutions, **(a)** and **(b)**.[1]

a) **The development of modified nucleosides aims to combat drug-resistant HIV mutants and potentially inhibit the emergence of such resistant strains.**

Clinical nucleoside reverse transcriptase inhibitors (NRTIs) are generally classified under the 2',3'-dideoxynucleoside (ddN) category, as illustrated in Figures 2 and 3. The absence of a 3'-OH group in ddNs is traditionally viewed as crucial for their role in terminating the DNA chain elongation process by reverse transcriptase (RT). Despite this, HIV mutants resistant to ddN-based drugs have rapidly and frequently emerged. This resistance suggests that these HIV variants can differentiate between ddNs and natural 2'-deoxynucleosides (dNs), rejecting ddN drugs from the active site of RT. To prevent the development of drug-resistant HIV mutants, modified nucleosides must meet two criteria:

1. The modified nucleosides should closely resemble dNs to avoid detection and rejection by HIV, making RT mistakenly incorporate them instead of natural dNs. This similarity necessitates the inclusion of a 3'-OH group in the modified nucleosides.
2. Despite possessing a 3'-OH group, these nucleosides must still function as effective terminators of RT-mediated synthesis of proviral DNA.

In line with these requirements, 4'-C-substituted-2'-deoxynucleoside (4SdN) (see Figure 3) was designed. This nucleoside fulfills both criteria by:

1. Being indistinguishable from dNs to HIV due to its comprehensive mimicry of dN's functional groups.
2. Featuring a 4'-substituent that renders the 3'-OH group a much less reactive, neopentyl-type secondary alcohol. While this modification allows HIV to mistake 4'SdN for a dN, it prevents the 3'-OH from participating in the elongation of the proviral DNA chain, thereby acting as a terminator of proviral DNA synthesis.

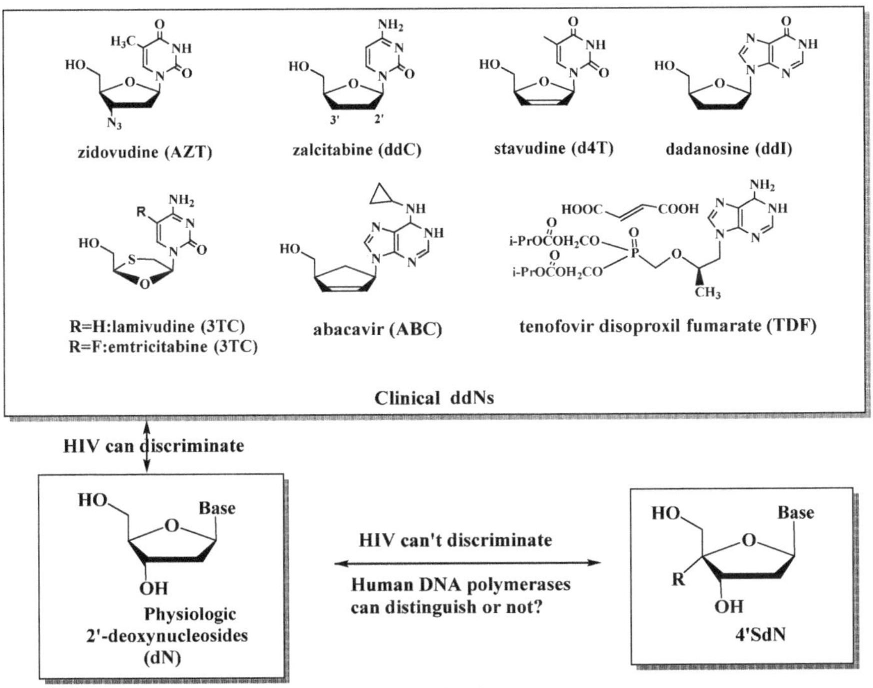

Fig. 3

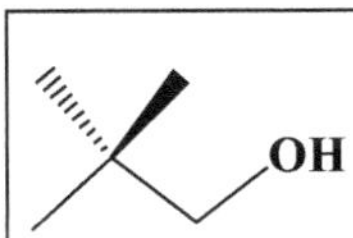

Neopentyl alcohol is a primary alcohol, but the hydroxyu is not reactive because the hydroxymethyl group is bonded to a tertiary carbon.
4'SdN has two neopentyl hydroxy groups, **5'-OH** (primary) and **3'-OH** (secondary).

It's important to recognize a potential challenge with 4'SdN due to its structure, which features a less reactive neopentyl-type primary alcohol at the 5'-position. A key concern is whether this unreactive 5'-OH group can undergo phosphorylation by kinase, an aspect that will be clarified by future research findings.

Additionally, the interaction between the 3'-OH group and the 4'-substituent in 4'SdN induces a preference for the furanose ring to adopt a 3'-endo conformation (N-type). This structural adjustment renders 4'SdN more resistant to degradation through acidic and enzymatic glycolysis compared to dN and ddN. During glycolysis, the oxygen atom of the furanose ring typically contributes to the formation of a coplanar oxocarbenium ion. However, the altered conformation in 4'SdN makes it challenging for this oxygen atom to achieve the coplanar arrangement necessary for glycolysis, as illustrated in Figure 4.

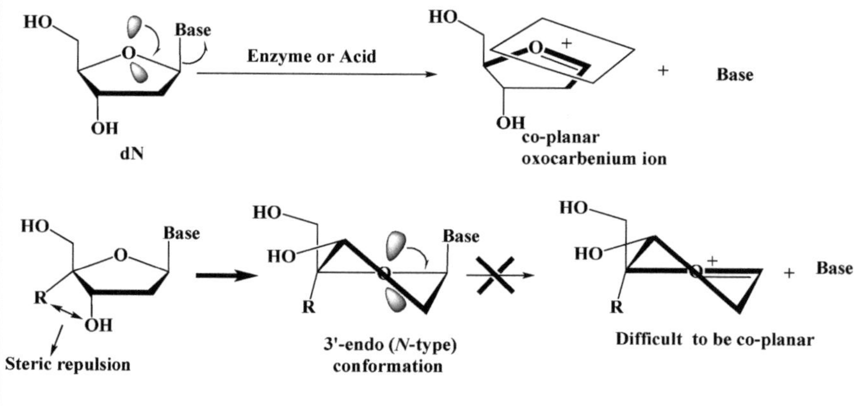

Fig. 4

The breakdown of nucleosides not only diminishes their biological effectiveness but also introduces new toxic effects through the release of bases. A case in point involves the anti-herpes drug Sorivudine, which led to fatalities among cancer patients concurrently treated with 5-fluorouracil (5FU). This was due to the release of 5-bromovinyluracil, which blocked the breakdown of toxic 5FU, resulting in death from its unmitigated toxicity (see Figure 5). This incident served as a cautionary tale for Yamasa Corporation, the developer of Sorivudine, making the company apprehensive about conducting clinical trials for EFdA due to their past adverse experiences in antiviral drug development.

1. Furthermore, the presence of an electron-withdrawing 3'-OH group in 4'SdN enhances its stability in acidic conditions compared to ddN, including stability against degradation by purines. This feature opens up possibilities for creating a variety of purine derivatives.

2. Additionally, the introduction of lipophilic substituents at the 4'-position increases the lipophilicity of 4'SdNs, improving their ability to cross cell membranes. This potentially boosts their bioavailability, making them more effective in reaching their target sites within cells.

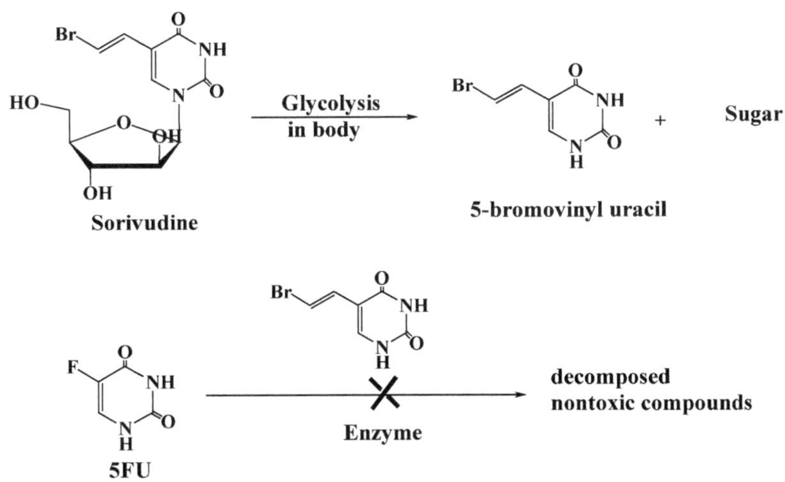

Several cancer patients who took 5FU died by taking Sorivudine due to the toxicity of the accumulated 5FU

Fig. 5

b) A strategy to reduce the toxicity of nucleoside drugs, based on my investigation into how the structure of nucleoside derivatives influences their biological activity, involves leveraging the differences in substrate recognition between reverse transcriptase (RT) and human DNA polymerases.

If human DNA polymerases incorrectly incorporate 4'SdN as they would natural deoxynucleosides (dNs), the consequences could be highly toxic. Despite this potential for toxicity, 2',3'-dideoxynucleosides (ddNs), which halt the action of DNA polymerases in a manner similar to the Sanger sequencing method and are thus inherently toxic, are effectively utilized as anti-HIV medications. This discrepancy arises because RT incorporates ddNs more readily than human DNA, indicating a significant difference in substrate discrimination between RT and human DNA polymerases.

This knowledge opens up the possibility of designing modified nucleosides that specifically target RT with greater selectivity, thereby reducing their toxicity to humans compared to current nucleoside reverse transcriptase inhibitors (NRTIs).

The structures of notable nucleoside antibiotics are depicted in Figure 6, highlighting the diversity of structures that could be optimized for selective activity against RT while minimizing adverse effects on human cells.

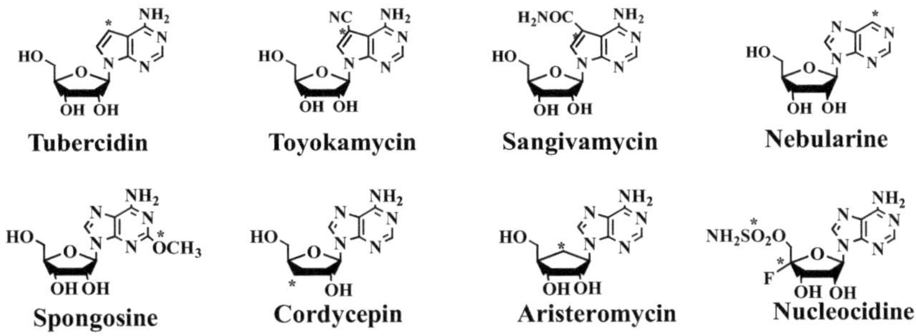

*modified position

Fig. 6

Most of these compounds are modified nucleoside derivatives, altered at a single point from their natural counterparts. While they exhibit strong activity against microorganisms, their high toxicity precludes their use in clinical settings. During the 1960s and 1970s, many organic chemists attempted to modify these antibiotics to enhance or introduce new biological activities. Unfortunately, further modifications often led to reduced or completely lost activity. Similarly, synthetic nucleosides altered at two or more points from natural nucleosides tended to be neither active nor toxic. This led many researchers in the field to believe that the future of nucleoside chemistry was bleak, given the challenge of creating derivatives with both high antibacterial effectiveness and low toxicity, leading some to abandon the field altogether.

However, these outcomes suggest that intercellular enzymes fail to recognize nucleoside derivatives modified at two or more positions as substrates, implying that while single-point modifications result in high activity and toxicity, multi-point modifications result in neither. This pattern indicates a potential pathway to reduce the toxicity of 4'SdNs, which are modified at a single position and thus likely to be highly toxic, through further modification.

The discovery of AZT as an anti-HIV agent by Dr. Mitsuya not only marked a significant breakthrough in HIV treatment but also revitalized interest in nucleoside chemistry, demonstrating that the field still held promise for developing effective and safer drugs.

Results and Discussion:

Assessing the validity of the proposed hypotheses using 4'-C-methyl-2'-deoxynucleosides

To assess the credibility of the proposed hypotheses, an initial step involved the synthesis and biological evaluation of various 4'-C-methyl nucleoside derivatives: 4'-C-methyl-D-ribonucleosides (MNs), 4'-C-methyl-2'-deoxy-D-nucleosides (MdNs), 4'-C-methyl-2',3'-dideoxy-D-nucleosides (MddNs), and 4'-C-methyl-2',3'-didehydrodideoxy-D-nucleosides (Md4Ns). The structural details of these 4'-C-methylnucleosides are illustrated in Scheme 1. It's important to note that since each of these nucleosides features a methyl group at the 4' position, the nomenclature for these compounds omits the specific mention of "4'" in their abbreviated names.

Synthesis of 4'-C-methyl nucleoside derivatives:

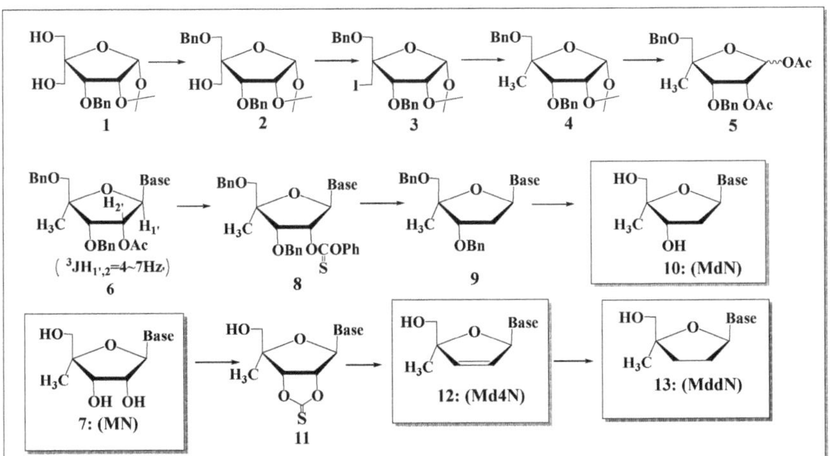

Scheme 1

The synthesis of 4'-C-methyl nucleoside derivatives commenced from diol 1, which was prepared via Moffatt's Aldol Condensation. The process involved selective benzylation of 1 to yield compound 2 with moderate success. Following this, compound 2 underwent a two-step transformation to convert its hydroxy group into a methyl group, resulting in compound 4. This was then converted into the diacetate 5. Subsequent glycosylation of 5 with various nucleobases produced compound 6, which upon deprotection, yielded 7 (MN). The removal of the 2'-OH group from 6 led to compound 9, which after deprotection became 10 (MdN). Additionally, the removal of both 2'-OH and 3'-OH from 7 resulted in compound 12 (Md4N), which upon catalytic hydrogenation of its double bond, yielded 13 (MddN) as outlined in scheme 1.

The deoxygenation process for the primary alcohol of 2 proved challenging due to its neopentyl-like structure, making conventional reduction methods ineffective. However, employing Garef's protocol with imidazole, triphenylphosphine, and iodine in a toluene and acetonitrile mixture at reflux successfully produced iodide 3. Barton Radical Deoxygenation of 3 then efficiently yielded compound 4.

It is worth noting that during the synthesis from compound 8 (with adenine base) to 9, the use of an excess of phenoxy thiocarbonyl chloride led to a deaminated product (with purine base) in a 12% yield. Moreover, the initial benzylation of diol 1 to 2 yielded moderate success and required chromatographic purification. This suggests a need for improved methodologies for the synthesis of compound 2. Additionally, compound 5 displayed a slower reaction time in glycosylation with silylated nucleobases compared to typical ribo-furanose derivatives, indicating its lower reactivity. Furthermore, the coupling constants ($^3J_{1',2'}$) of compound 6 were notably higher than those observed in standard β-D-ribo-nucleosides, suggesting a conformational change in the furanose ring of 6.

The evaluation of anti-HIV activity (EC_{50}) and toxicity (CC_{50}) for these 4'-C-methylnucleosides revealed that MdN exhibited significant biological activity and toxicity, whereas MddN and Md4N showed minimal activity. These findings underscore the critical role of the 3'-OH group for biological activity and suggest that while the neopentyl-type 5'-OH in MdN might be phosphorylated by kinase, those in MddN and Md4N were less likely to undergo phosphorylation. This phosphorylation is vital for their biological function, as nucleosides exert their biological effects in the form of their 5'-O-triphosphate. Additionally, studies have shown that the 5'-O-triphosphate forms of both 4'-C-methyl-2'-deoxycytidine (MdCTP) and 4'-C-methyl-D-arabinofuranosyl cytidine (MaraCTP) act as terminators for calf thymus DNA-polymerase α and recombinant rat DNA-polymerase β, indicating that 4'SdN functions as an NRTI. Despite these insights, further research on MdCTP with RT was not conducted. Lastly, the 4'-C-methyl-D-ribofuranosyl nucleosides (MNs) displayed no biological activity, likely due to the inability of their 5'-OH groups to be phosphorylated by kinase.

Table 1. Anti-HIV activity of 4'-*C*-methyl-2'-deoxynucleosides

Structure	Base	EC_{50}(μM)	CC_{50}(μM)	SI
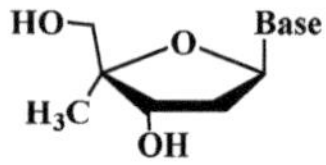 MdN	Ad	2.6	2.6	1.0
	Th	7.2	104	14.4
	Cy	0.072	0.13	1.8
	Pu	1.9	>200	105
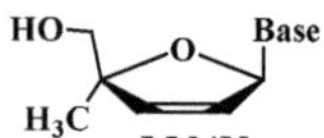 Md4N	Ad	>500	>500	~1
	Th	21	330	15.7
	Cy	350	350	1
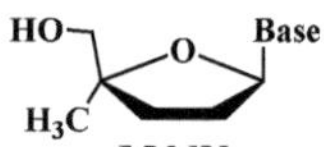 MddN	Ad	30	400	13.3
	Th	>500	>500	~1
	Cy	27	27	1
AZT		0.001	>20	>2020
ddA		47	>500	>11
d4T		4.1	>500	>120

EC_{50}= 50% effective concentration: **CC_{50}**= Cytotoxic concentration: **SI**=selectivity index (**CC_{50}/EC_{50}**)
MdN=4'-*C*-methyl-2'-dexynucleosides: **Md4N**=4'-*C*-methyl-2',3'-didehydrodideoxynucleoside;
MddN=4'-*C*-methyl-2',3'-dideoxynucleoside: **AZT**=3'-azido-3'-deoxythymidine:
ddA=2',3'-dideoxy adenosine: **d4T**=2',3'-didehydro-3'-deoxythynidine.

.

Concerning Moffatt's Aldol Condensation:

During my tenure as a post-doctoral fellow under Dr. Moffatt at Syntex Research between 1973 and 1974, Dr. Youssefyeh undertook an experiment to perform an aldol condensation using D-*ribo*-aldehyde 14 and formaldehyde, aiming to synthesize a 4'-*C*-hydroxymethyl-D-ribose derivative, designated as compound 15. Contrary to expectations, this reaction resulted in an equal mix of two distinct products, each characterized by a 1,2-*O*-isopropylidene group and three hydroxyl groups, respectively. The exact structures of these products remained undetermined by his team. In response, I proposed a potential mechanism that might explain the formation of these products through a combination of retro-aldol and aldol condensation reactions, as depicted in Scheme 2.

To mitigate the occurrence of retro-aldol condensation as outlined in Scheme 2, the research approach was adjusted by substituting the initial reactant, compound 14, with its 3-*O*-benzyl derivative. This modification aimed at stabilizing the reaction and steering it towards the desired product formation.

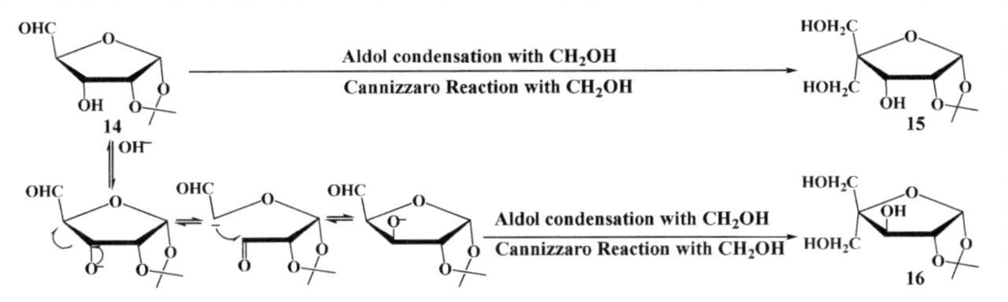

Scheme 2

I am deeply familiar with Moffatt's Aldol Condensation and applied it in my research endeavors. Prior to initiating my investigation into 4'SdN, I reached out to Dr. Moffatt with an update on my health and my intentions to explore the development of anti-HIV modified nucleosides using his aldol condensation technique. In my message, I shared updates about undergoing surgeries for both atrial septal defect (ASD) and a stroke, which was a complication of the heart surgery, at the age of 42.

By then, Dr. Moffatt had retired from Syntex Research, and in his stead, I received several of his reprints concerning 4'-C-substituted nucleosides from another party. Realizing the limited research in this area, I embarked on my study of 4'SdN, guided by my theoretical frameworks. In addition, I later innovated a novel aldol condensation method for synthesizing 4'-C-substituted D-ribo-furanoses, utilizing formaldehyde and 1,2-O-isopropylidene-3-keto-furanose, a technique I will elaborate on further.

Exploring the Structure-Activity Relationship (SAR) of 4'SdNs:

Subsequently, to investigate the structure-activity relationship (SAR) of 4'SdNs and to develop versions with enhanced anti-HIV efficacy and reduced toxicity compared to MdNs, we synthesized 4'SdNs featuring a variety of 4'-C-substituents and nucleobases. These compounds were then assessed for their biological activity. During the course of our research, the anti-HIV properties of 4'SdNs were also documented by the Syntex research team and other researchers. As a result, the anti-HIV activities of the 4'SdNs analyzed in our study, alongside those identified by other groups, are collectively presented in Table 2.

Table 2. Anti-HIV activity of 4'-*C*-substituted-2'-deoxynucleosides

Compounds	$EC_{50}(\mu M)$[a]	$CC_{50}(\mu M)$	S.I.
4'-*C*-cyanothymidine	0.002	1	500
4'-*C*-azidothymidine	0.01	8	300
4'-*C*-ethynylthymidine	0.83	>400	>482
4'-*C*-ethynylarabinofuranosylthymidine	119	>400	159
4'-*C*-azidomethylthymidine	2.1	33.3	14
4'-*C*-methylthymidine	7.2	104	14.5
4'-*C*-ethylthymidine	>400	400	ND
4'-*C*-methoxythymidine	8.49	200	24
4'-*C*-vinylthymidine	>400	>400	ND
4'-*C*-hydroxymethylthymidine	7.0	>400	ND
4'-*C*-propylthymidine	>100	>100	ND
4'-*C*-cyano-2'-deoxycytidine	0.0012	0.17	141.7
4'-*C*-azido-2'-deoxycytidine	0.004	0.21	52
4'-*C*-ethynyl-2'-deoxycytidine	0.0048	2.2	52
L-4'-*C*-ethynyl-2'-deoxycytidine	>400	>400	ND
4'-*C*-ethynyl-2'-deoxy-5-fluorocytidine	0.030	>100	>333
4'-*C*-methynylarbinofuranosyl-cytidine	0.043	2.0	46.5
4'-*C*-methyl-2'-deoxycytidine	0.015	1.0	66.7
4'-*C*-fluoromethyl-2'-deoxycytidine	68	0.12	0.12
4'-*C*-methyl-2'-deoxyadenosine	2.6	2.6	1
4'-*C*-azido-2'-deoxyadenosine	0.13	50	385
4'-*C*-ethynyl-2'-deoxyadenosine	0.098	16	1630
2',3'-dideoxy3'-thia-L-cytidine (**3TC**)	0.10	>100	>1000
3'-azido-3'-deoxythymidine (**AZT**)	0.0032	29.4	9190

[a] Anti-HIV activity was determined by MTT assay. MT-4 cells and HIV-1_{LAI} were employed. ND: not determined.

1. The potency of anti-HIV activity in relation to 4'-substituents appears to follow this order: ethynyl (C≡CH) is more effective than N3, which is more effective than CH=CH2, with Me and Et being similarly effective, and

C≡C-CH3 being the least effective. Interestingly, this sequence is opposite to the energy difference (-ΔG0) observed for substituents positioned equatorially versus axially on a cyclohexane ring, where ethynyl (C≡CH) is less demanding sterically than others. This suggests that substituents at the 4'-position that are less bulky enhance anti-HIV activity. It was later discovered that the ethynyl and CN groups at the 4'-position of 4'SdN form strong hydrophobic interactions with RT by fitting into a small hydrophobic pocket of RT.

2. Purine analogs typically exhibit lower toxicity than pyrimidine analogs. For instance, 4'-C-ethynyl-2'-deoxy-5-fluorocytidine, a dual-modified nucleoside derivative of 2'-deoxycytidine, showed a favorable Selectivity Index (SI = CC_{50}/EC_{50}) in MT-4 cells but was found to be toxic in other cell types, according to a private communication from Dr. Kohgo of Yamasa.

3. Arabino analogs generally display lower toxicity than their 2'-deoxy counterparts.

4. Both 4'SddNs and 4'Sd4Ns lack significant anti-HIV activity, potentially because their neopentyl-type 5'-OH group cannot be phosphorylated by kinases.

5. The L-isomers of 4'SdN exhibit no anti-HIV activity, despite the known activity and lower toxicity of the L-enantiomer of 2',3'-dideoxy-3'-thia-L-cytidine (3TC) compared to its D-enantiomer. This lack of activity in L-isomers of 4'SdNs might be attributed to excessive modification, rendering them unrecognizable by RT as substrates.

Additionally, two 4'-C-methyl analogs of Sorivudine, 4'-C-methyl-2'-deoxy-D-ribofuranosyl-5-bromouracil (MBVU) and 4'-C-methyl-D-arabinofuranosyl-5-bromouracil (MBVaraU), were synthesized and assessed for their anti-herpes activity. The outcomes of their activity and cytotoxicity are detailed in Table 3.

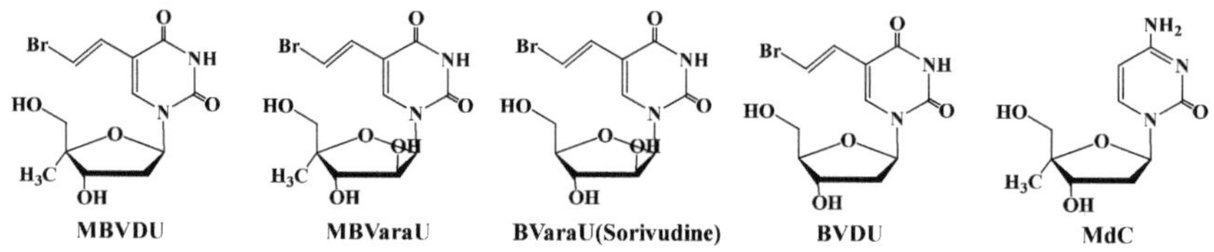

Fig. 7

Table 3. Antiviral activity and Cytotoxicity of 4'-*C*-Methylnucleosides

compound	Antiviral Activity ED50(μg/mL)		Cytotoxicity ID50(μg/mL)	
	HSV-1 [a,d]	HSV-1 [b,d]	HSV-1 [c,d]	CCRF-HSB-2 [e]
MBVDU	0.0053	0.26	0.00077	0.45
MBVaraU	24.4	63.5	0.18	100>
MdC	0.071	0.27	0.094	0.12
BVDU	0.052	100>	0.013	100>
BVaraU (Sorivudine)	0.048	62	0.00083	100>

[a]HSV-2 -1 VR-3strain, [b]HSV-2 MS strain, [c]VZV Oka strain, [d]plaque reduction assay, [e]MTT assay

While MBVDU exhibits higher activity against the Herpes Virus than Sorivudine, its toxicity is a significant concern. Given that these nucleosides act by inhibiting DNA polymerases in their triphosphate forms, the variation in their activity and toxicity levels can be understood as follows: the 5-OH group in BVDU and BVaraU is phosphorylated by viral kinases similar to acyclovir, rendering them effective against herpes without being toxic. Conversely, MBVDU and MdC undergo phosphorylation by cellular kinases, making them both active against herpes and toxic. Meanwhile, MBVaraU is scarcely phosphorylated by either viral or cellular kinases, leading to its lower activity and toxicity.

Creation of purine derivatives from 4'-C-cyano-2'-deoxynucleosides (CNdNs) and 4'-C-ethynyl-2'-deoxynucleosides (EdNs) and the evaluation of their anti-HIV effectiveness:

The findings from our research prompted an investigation into the biological activity of purine derivatives from CNdNs and EdNs. While our project's 4'SdNs were synthesized through the glycosylation of 4'-C-substituted-D-ribofuranose derivatives with nucleobases, we encountered several challenges with this synthetic approach:

1. The synthesis of 4'-*C*-substituted-D-ribofuranoses and their subsequent transformation into 4'SdNs involved multiple steps, leading to low overall yields.
2. The 4'-C-Substituted-D-*ribo*-furanose derivatives exhibited poor reactivity in glycosylation reactions, particularly when the substituent was an electron-withdrawing group, such as a cyano group.

Moreover, the reduced reactivity at the anomeric position of these 4'-C-substituted furanose derivatives suggested that 4'SdNs would exhibit enhanced stability against both acidic and enzymatic hydrolysis compared to dNs and ddNs.

These challenges led us to explore an alternative synthetic route for creating purine derivatives of 4'SdN, beginning with dNs (Scheme 3).

Scheme 3. Synthetic scheme of purine derivatives of 4'-C-ethynyl-2'-deoxynucleoside and 4'-C-cyano-2'-deoxynucleoside from 2'-deoxynucleosides.

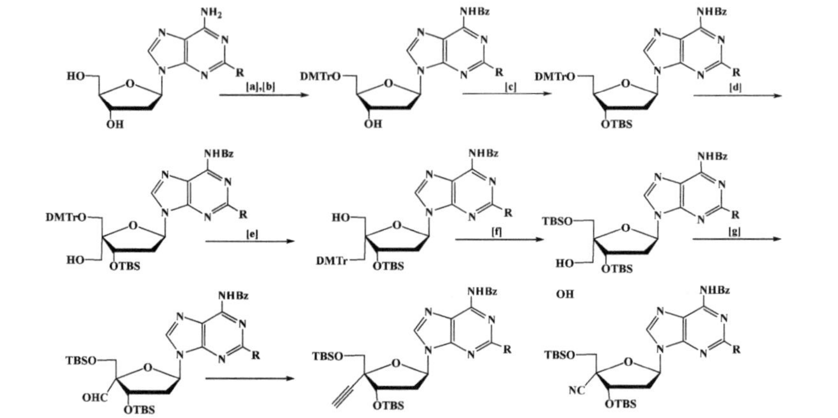

[a]:1.TMSCl,py,0°C; 2.BzCl,py,0°C; 3.NH_4OH,H_2O,0°C, [b]:DMTrCl,py,rt, [c]:1.TBSCl,imidazole,DMF,rt;2.TsOH,MeOH,$CHCl_3$,0°C,
[d]:1.EDC-HCl,py,TFA,toluene,DMSO,rt; 2.aq.CH_2O,1N NaOH,dioxane,rt; 3.$NaBH_4$,EtOH,0°C, [e]: DMTrCl,CH_2Cl_2,0°C,
[f]:1.TBSCl,imidazole,DMF,rt; 2.TsOH,MeOH,$CHCl_3$,rt, [g]:EDC-HCl,TFA,toluene,DMSO,rt.
TFA=trifluoroacetic acid, DMF=dimethylformamide, rt=room temperature

The biological activity of these compounds is presented in Table 4 and can be summarized as follows:

1. Several purine derivatives of CNdN exhibit potent anti-HIV activity, although none achieve a satisfactory Selectivity Index (SI).
2. All purine derivatives of EdN demonstrate strong anti-HIV activity along with acceptable levels of SI.

Table 4. Anti-HIV activity of purine derivatives of 4'-*C*-cyano-2'-deoxynucleoside (CNdN) and 4'-*C*-ethynyl-2'-deoxynucleoside (EdN).

structure	Base	$EC_{50}(\mu M)^{a)}$	$CC_{50}(\mu M)$	SI
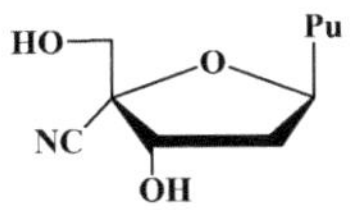	A	0.051	12	235
	I	0.051	23	451
	2AA[b)]	0.00079	0.034	43
	G	0.000188	0.034	181
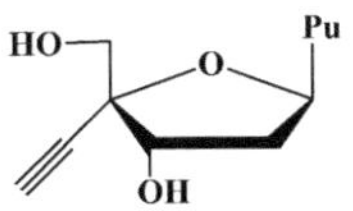	A	0.098	16	1630
	I	0.15	216	1440
	2AA[b)]	0.0003	0.82	2733
	G	0.0014	1.5	975
AZT		0.0032	29.4	9190

[a)] Anti-HIV activity was determined by MTT assay. MT-4 cells and HIV-1 $_{LAI}$ were employed.
A=adenine, I=hypoxanthine, 2AA=2-amiboadenine, G=guanine

Efficacy of 4'SdNs in Targeting Drug-Resistant HIV Strains

[20, 24, 25, 28, 29]

Numerous 4'SdNs have demonstrated exceptional anti-HIV efficacy against the wild-type HIV. However, a critical aspect of my research focuses on their effectiveness against HIV strains that have developed resistance to existing NRTI drugs. The efficacy of select 4'SdNs against various NRTI-resistant HIV strains is detailed in Table 5.

It is particularly significant that three 2'-deoxycytidine derivatives retained their effectiveness against these drug-resistant HIV mutants, though the activity of 4'-C-ethynyl-D-arabinofuranosyl cytidine (EaraC) and MdC was notably reduced against mutations M184V, M184I, and the multi-mutation strains 41/69/125/SG. In contrast, the purine derivatives, specifically 4'-C-ethynyl-2'-deoxyadenosine (EdA), 4'-C-ethynyl-2'-deoxy-2-aminoadenosine (Ed2AA), and 4'-C-ethynyl-2'-deoxyguanosine (EdG) — with the exception of 4'-C-ethynyl-2'-deoxyinosine (EdI) — demonstrated substantial potency against all examined drug-resistant

HIV mutants. Notably, EdI exhibited significantly lower activity compared to the other three derivatives, especially against the M184V mutation.

Table 5. Anti-HIV activity of selected 4'-C-substituted-2'-deoxynucleoside against wild type HIV and drug-resistant HIVs

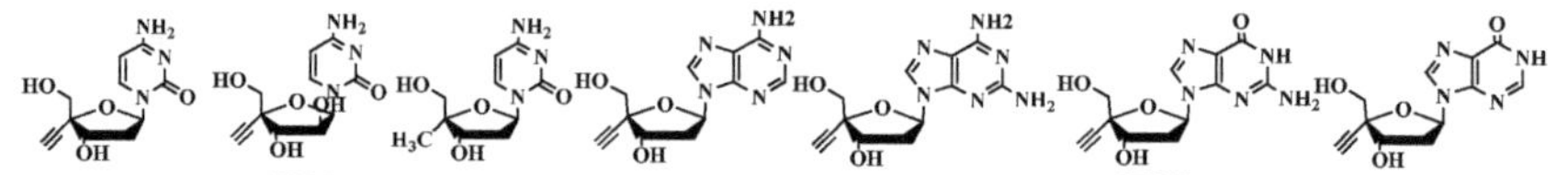

4'EdC **4'EdC** **4'MdC** **4'EdA** **4'Ed2AA** **4'EdG** **4'EdI**

Compound	EC50(μM)[a)									
	HXB2[b)]	KH65R	L74	41/215	M184V	M184I	41/69/ 125/SG	MDR[c)]	Y181C	CC_{50}(μM)
4'EdC	0.0012	0.0008	0.0013	0.006	0.0024	0.0026	0.015	0.0012	0.0021	>200
4'EaraC	0.0071	0.015	0.026	0.026	0.71	0.48	0.17	0.0079	0.0016	>200
4'MedC	0.0058	0.0071	0.0062	ND	0.2	0.74	ND	0.0033	ND	>200
4'EdA	0.008	0.0033	0.004	0.012	0.047	0.022	0.065	0.0062	0.011	>200
4'Ed2AA	0.0014	0.00035	0.007	0.0017	0.0059	0.0027	0.0041	0.001	0.0008	>200
4'EdG	0.007	0.001	0.0012	0.019	0.008	0.0041	0.0068	0.0048	0.01	52
4'EdI	0.81	0.25	0.61	1.3	1.6	1.5	2.2	0.51	ND	>200
AZT	0.022	0.02	0.02	0.3	0.001	0.017	1.6	15.3	0.0014	>100
3TC	0.71	ND	ND	ND	>100	>100	9.9	1.1	ND	>100
ddC	0.2	3.0	1.5	ND	2.2	ND	1.3	5.5	ND	>100
ddI	3.9	12.7	19.5	3.6	10.1	ND	12.2	25	ND	>100

[a)]Anti-HIV activity was determine with MAGI assay, ND: not determined, [b)]wild type HIV. [c)]multidrug-resistant HIV.

Moreover, the three purine derivatives demonstrated effectiveness against the Y181C mutation, which is resistant to non-nucleoside reverse transcriptase inhibitors (NNRTIs). These compounds also showed high potency against HIV strains from seven patients with advanced AIDS who had extensive drug treatment histories, performing as well against these strains as they did against the wild-type HIV. Consequently, EdA, Ed2AA, and EdG proved to be highly effective against all tested HIV variants. These findings suggest that these three purine EdNs might even have the potential to prevent the development of drug-resistant HIV strains.

It is important to highlight, however, that EdG exhibited toxicity to Hela cells at a concentration of 52μM, indicating it could be significantly toxic.

Toxicity of Purine EdNs in Mice

Given the high efficacy against all HIV strains and acceptable Selectivity Index (SI) of the three purine EdNs, their toxicity in mice was subsequently evaluated (as shown in Table 6). While eight mice managed to survive a single administration of EdA and EdI at doses ranging from 3 to 100mg/kg, delivered both intravenously and orally, all mice succumbed to a single 3mg/kg dose of Ed2AA and EdG, regardless of the method of administration. This outcome suggested that EdA and EdI were relatively non-toxic, whereas Ed2AA and EdG exhibited significant toxicity. Further investigation revealed that in mice, EdA and Ed2AA were rapidly converted into EdI and EdG, respectively, by the action of adenosine deaminase, indicating a metabolic pathway influencing their toxicity profiles.

Table 6. Toxicity of purine EdNs to mice[a)]

	Intravenous administration		oral administration	
	Dose(mg/Kg)	Morality (%)	Dose(mg/Kg)	Morality (%)
EdA	100	0	100	0
and	10	0	10	0
EdI	3	0	3	0
	1	0	1	0
Ed2AA	100	100 (1 day)[b)]	100	100 (1 day)
	10	100 (2 days)	10	100 (2 days)
	3	0	3	0
	1	0	1	0
EdG	100	100 (1 day)	100	100 (1 day)
	10	100 (2 days)	10	100 (4 days)
	3	100 (4 days)	3	100 (4 days)
	1	0	1	0

[a)] Six-week-old male were employed. [b)] Numbers in parenthrses represent survival days of mice after administration.

The findings indicate that accurately assessing the toxicity of EdA and Ed2AA in animals is challenging due to their susceptibility to metabolic processes.

Efficacy against HIV of EdA Derivatives Resistant to Adenosine Deaminase:

In light of the discovery that EdA and Ed2AA are readily deaminated by adenosine deaminase, efforts were directed towards developing EdA derivatives resistant to adenosine deaminase degradation. Previous research has shown that adenine derivatives substituted with a halogen atom at the 2-position of the base exhibit resistance to adenosine deaminase. Consequently, 4'-C-ethynyl-2'-deoxy-2-fluoroadenosine (EFdA) was synthesized as an initial candidate. This compound, which features modifications at both the 4' and 2-positions of the natural 2'-deoxyadenosine molecule, was evaluated for its anti-HIV efficacy, stability against adenosine deaminase, and resistance to acidic degradation. Given its dual modifications, EFdA's toxicity is anticipated to be lower than that of EdA. According to the data presented in Table 7, EFdA demonstrates significant potency against all HIV strains, including those resistant to multiple drugs and the M185V mutation, and boasts an exceptionally high Selectivity Index (SI) of 110,000.

Table 7. Anti-HIV activity of 4'-*C*-ethynyl-2-fluoro-2'-deoxyadenosine (EFdA)

HIV-1	Compound	EC50 (μM)	CC50 (μM)	SI
wild-type	Ed2AA	0.34	900	2,600
	EFdA	0.068	7,500	110,000
	AZT	3.2	29,400	9,100
M185V	EFdA	3.1	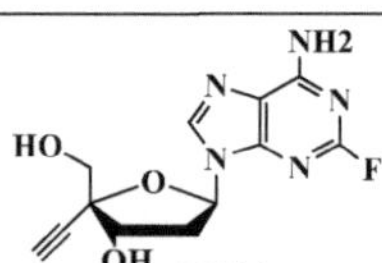 EFdA	m.p. (decomp) 224.4-224.6°C
	AZT	16.0		
MDR	EFdA	0.15		[α]D +4.64
	AZT	15,300.00		(c=0.5, DMSO)

CC_{50} = 50% effective concentration; CC_{50} = 50% cytotoxic concentration; SI = CC_{50}/EC_{50}

As anticipated, EFdA exhibited high stability to adenosine deaminase, remaining intact in conditions where EdA was fully deaminated within 60 minutes (see Figure 8). Additionally, EFdA showed considerable resistance to acidic conditions; only a minor fraction (3%) of EFdA degraded after 120 minutes in gastric juice (pH 1.06) at 24°C, in stark contrast to 2',3'-dideoxyadenosine (ddA), which was entirely decomposed under the same conditions (refer to Figure 8).

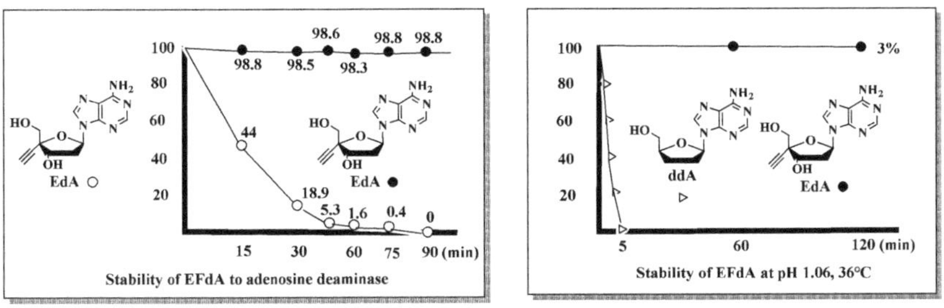

Stability of EFdA to adenosine deaminase

Stability of EFdA at pH 1.06, 36°C

Fig. 8

Publication of two studies attributing the toxicity of 4'SdNs to their 3'-OH group

During the course of our research, two publications emerged discussing the role of the 3'-OH group in 4'SdNs, suggesting that this group was responsible for the compounds' toxicity. Notably, one of the authors of these studies was Dr. Hiroaki Mitsuya, who, interestingly, is a collaborator on our project. Dr. Hiromichi Tanaka and his team presented findings that 4'-C-ethynyl-2',3'-didehydrodideoxythymine (Ed4T, illustrated in Figure 9) is approximately ten times more potent and less toxic compared to the clinically used d4T, positioning Ed4T as a highly promising candidate among anti-HIV modified nucleosides. They argued that the presence of the 3'-OH group in 4'SdNs significantly contributes to their toxicity.

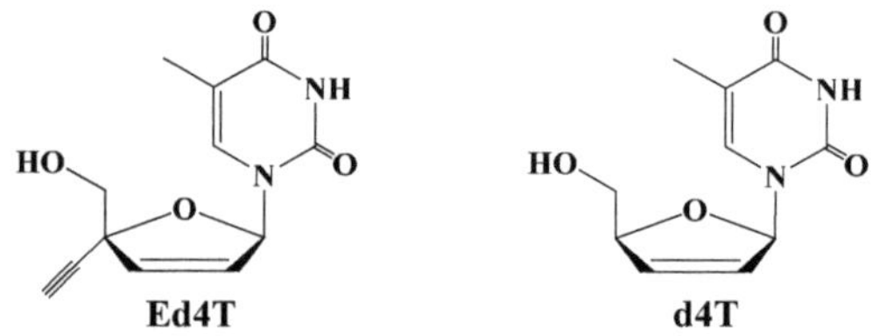

Fig. 9

While the two studies highlighted the toxicity attributed to the 3'-OH group in 4'SdNs, they did not address Ed4T's efficacy against drug-resistant HIV strains, a crucial aspect for advancing beyond the limitations of existing Highly Active Antiretroviral Therapy (HAART). Dr. Tanaka, well-acquainted with me, extended an apology for advancing the synthesis of Ed4T ahead of my work on 4'-*C*-ethynyl deoxy-nucleosides, knowing my research direction. I reassured him, emphasizing my focus on nucleosides with the 3'-OH group due to its significance in preventing drug-resistant HIV mutations. I proposed that Ed4T's reduced toxicity compared to d4T might stem from its additional modification and suggested its enhanced activity could be due to the ethynyl group's specific affinity for RT. This hypothesis was later confirmed through an X-ray study of Ed4T and RT complexes, revealing the ethynyl group's interaction with a hydrophobic pocket of RT. Subsequent research by Michailidis echoed this finding for EFdA a year later.

Conversely, Marquez and Mitsuya noted the 3'-OH group's role in 5'-OH phosphorylation by cellular kinases but also associated it with the toxicity of 4'SdNs. They found that while EddC was ineffective against HIV in cellular contexts, its triphosphate form (EddCTP) outperformed AZTTP in vitro, yet was less effective against the RT of the M184V mutant.

At a virology conference in Kumamoto, Japan, when questioned about EFdA's low toxicity, Mr. Tanaka credited the design of dual-modified nucleosides. Later, Dr. Mitsuya inquired via email about EFdA's non-toxicity, to which I explained it was due to the differential substrate selectivity between RT and DNA-polymerases; RT easily incorporates EFdA, whereas DNA-polymerases do not, showcasing the potential of various modified nucleosides as antiviral drugs.

This study provides a summary of the anti-HIV activity of 4'-C-substituted nucleosides (4'SNs), as illustrated in Figure 9.

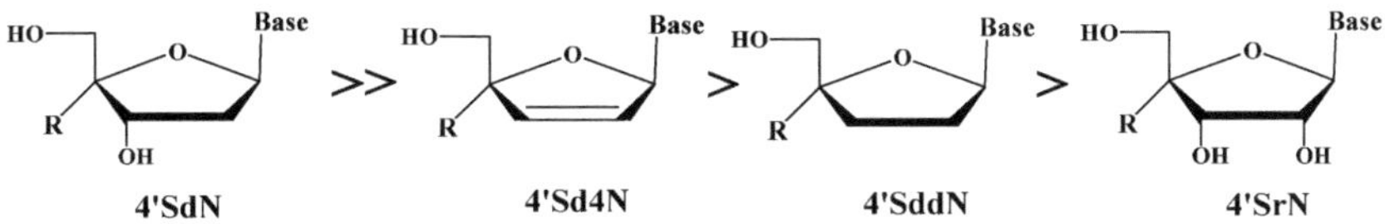

Fig.9

1. 4'-C-Substituted-D-ribofuranosyl nucleosides (4'SrN) exhibited no biological activity due to their neopentyl-type 5'-OH group's inability to be phosphorylated by cellular kinases, though chemically synthesized 4'SrN-5'-O-monophosphates showed potential anti-tumor effects.

2. The anti-HIV efficacy of 4'SddNs was lower than their ddN counterparts because their neopentyl-type 5'-OH group was more difficult for cellular kinases to phosphorylate.

3. Similarly, 4'Sd4Ts, with the exception of Ed4T, had reduced anti-HIV activity compared to d4Ts, for the same phosphorylation challenges. (Ed4T, however, was ten times more effective than d4T due to its 4'-C-ethynyl group's unique affinity for RT.)

4. 4'SdNs demonstrated significant anti-HIV activity, as their neopentyl-type 5'-OH could be phosphorylated by cellular kinases, aided by the presence of a 3'-OH group.

Following the publication of these insights, a representative from Yamasa Corporation's R&D division approached me with concerns about not having synthesized Ed4T before Dr. Tanaka, highlighting Ed4T's promise as an anti-HIV agent and expressing a desire to end our collaboration due to the perceived high toxicity of nucleosides with a 3'-OH group. They also wished to secure patents for all my modified nucleosides and requested confidentiality on 4'SdN data until patent filings were complete. Despite EFdA's synthesis and possession by Yamasa, its biological assessment had been delayed over toxicity concerns.

I explained my focus was on nucleosides with a 3'-OH group, showing less interest in those without, and agreed to their request regarding data confidentiality. I also recommended assessing EFdA's biological efficacy, considering its dual modifications might result in lower toxicity. Consequently, all patent rights for EFdA were assigned to Yamasa Corporation.

Despite these developments, I pursued my research on EFdA, confident in its reduced toxicity due to the dual modifications. EFdA's exceptional anti-HIV activity later became widely recognized, with Dr. Mitsuya presenting its development globally and discussing its design strategy.

Efficacy Against HIV of 2',3'-Dideoxy (dd) and 2',3'-Didehydrodideoxy (d4) Analogs of EFdA

The publication of the aforementioned studies inspired me to create dd- and d4-analogs of EFdA and assess their effectiveness against HIV. The anti-HIV activities of 4'-C-ethynyl-2-fluoro-2',3'-dideoxyadenosine (EFddA), 4'-C-ethynyl-2-fluoro-2',3'-didehydrodideoxyadenosine (Ed4A), and 4'-C-ethynyl-2-chloro-2'-deoxyadenosine (ECldA), in comparison with d4T, Ed4T, 3TC, and AZT, are detailed in Table 8.

Table 8. The anti-HIV activity of 4'-*C*-ethynyl-2'-deoxynucleosides

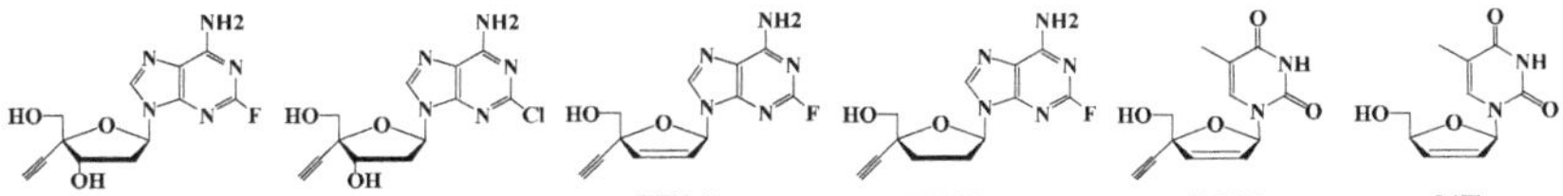

EFdA ECldA EFd4A EFddA Ed4T d4T

Compound	Anti-HIV activity (Magi assay, μg)			
	$HIV\text{-}1_{wild}$	$HIV\text{-}1_{MDR}$	$HIV\text{-}1_{184V}$	SI
EFdA	0.00020	0.00014	0.0044	110,000
ECldA	0.0019	0.0084	0.01	330,000
EFd4A	0.80	0.15	1.8	
EFddA	0.94	8.7	97	
AZT	0.17	74.3	0.13	
3TC	1.0	2.8	>100	
Ed4T	1.5	1.1	17	>50,000
d4T	7.6	64	5.6	

The findings highlight the critical role of the 3'-OH group in combating drug-resistant HIV strains. To date, the HIV mutant demonstrating the highest resistance to EFdA has been identified as M184V/T165R/142, exhibiting a resistance level 22 times greater than that of the wild-type HIV. Nonetheless, EFdA remains effectively active against this mutant, successfully hindering the development of further resistant strains over the past two decades.

This groundbreaking discovery underscored the distinct substrate preferences between RT and DNA polymerases, with RT favoring the uniquely modified EFdA. The enhanced activity of Ed4T and the superior substrate efficiency of EFdATP for RT further underscored the special affinity of the 4'-C-ethynyl group for RT. X-ray crystallography studies confirmed this affinity, revealing the precise interaction of the ethynyl group with a specific hydrophobic pocket on RT.

Despite these successes, the development of Ed4T was halted due to the emergence of resistant HIV mutants, attributed to the absence of a 3'-OH group which complicated its phosphorylation process.

Toxicity of EFdA in Mice and its Effects on Human DNA Polymerases

Given EFdA's resistance to adenosine deaminase and its potent efficacy against all HIV strains, its toxicity in mice was investigated. EFdA exhibited no acute toxicity in mice at doses up to 100mg/kg, regardless of whether it was administered orally or intravenously. Additionally, changes in the body weight of mice following EFdA administration indicated no toxic effects of EFdA on mice.

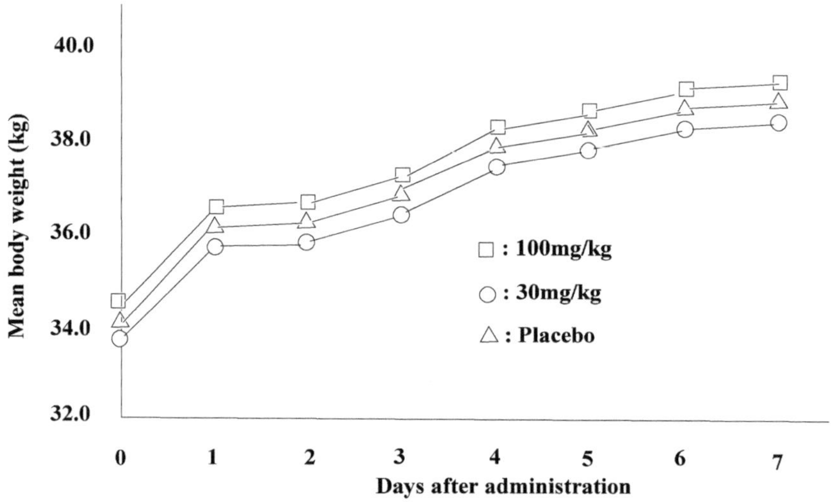

Fig. 10 Body weight change of mice after a single dosage of EFdA administrated orally or intravenously to ICR mice.

The toxicity associated with nucleoside reverse transcriptase inhibitors (NRTIs) is primarily due to their ability to inhibit mitochondrial DNA polymerase γ. The concentration at which 4'-C-ethynyl-2'-deoxy-2-fluoroadenosine-5'-O-triphosphate (EFdATP) inhibits 50% of 2'-deoxyadenosine-5'-O-triphosphate (dATP) incorporation by human mitochondrial DNA polymerase γ is 10μM. This value is significantly higher compared to the 0.2μM observed for 2',3'-dideoxyadenosine-5'-O-triphosphate (ddATP), indicating a reduced potential for mitochondrial toxicity. Furthermore, the inhibitory effects of EFdATP on DNA polymerase α and β are minimal, with EC_{50} values exceeding 200μM, suggesting that these human DNA polymerases hardly recognize EFdATP, a molecule modified at two positions from the natural dATP, as a substrate, while reverse transcriptase (RT) does recognize it.

Additionally, EFdATP has been found to be twice as effective a substrate for RT compared to the natural substrate dATP, underscoring its efficiency in targeting HIV. EFdA has also demonstrated significant activity against Simian Immunodeficiency Virus (SIV) without any detectable adverse effects in macaques over six months of continuous treatment, highlighting its potential as a safe and effective therapeutic option.

Table 9. Toxicity of EFdA and EFdATP and stability of EFdATP in plasma

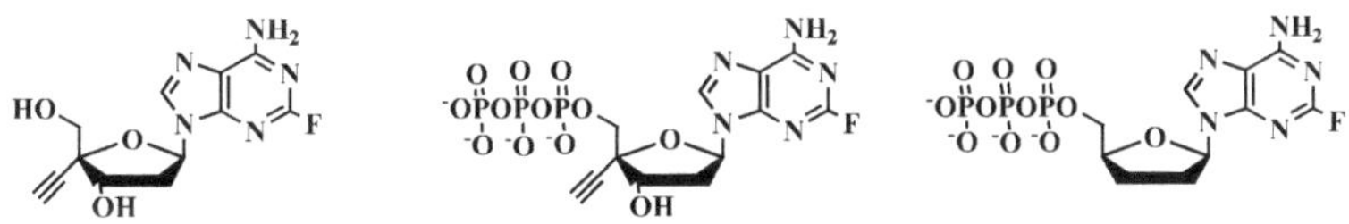

EFdA **EFdATP** **ddATP**

Mouse Toxicity of EFdA: No acute toxicity up to 100mg/kg

Inhibition of DNA-plolymerase by EFdATP (IC_{50})	
DNA-polymerse α	>200μM
DNA-polymerse β	>200μM
Mitocondrial DNA-polymerse γ	10μM
ddATP	0.2μM

Half life ($T_{1/2}$) of EFdATP in plasma: ~18h

AZTTP: ~3h

Intracellular Metabolism of EFdA [40]

With an increase in the concentration of intracellular EFdA, there was a proportional increase in all its metabolite fractions, including EFdA-monophosphate (EFdAMP), EFdA-diphosphate (EFdADP), and EFdA-triphosphate (EFdATP). This pattern contrasts with that observed for AZT metabolites, where only the monophosphate form (AZTMP) showed a significant increase in response to higher intracellular AZT levels, while AZT-diphosphate (AZTDP) and AZT-triphosphate (AZTTP) did not demonstrate similar proportional increases.

The intracellular half-life (T1/2) of EFdATP was approximately 18 hours in various cell types including CEM cells, MT4 cells, and MAGI-CCR5 cells, significantly longer than the half-life of AZT, which was about 3 hours. Notably, around 50% of MT4 and MAGI cells remained protected from HIV infection for 24 hours after the removal of extracellular EFdA, when cultured in the presence of 0.1 μM EFdA. This suggests that EFdADP and EFdATP, in particular, exhibit remarkable stability against breakdown by intracellular enzymes. These findings underscore the exceptional intracellular stability of EFdA and its metabolites, EFdADP and EFdATP, against enzymatic degradation, highlighting their potential effectiveness in HIV therapy.

A rationalization of the inhibition of RT and DNA-polymerases by 4'SdNs.

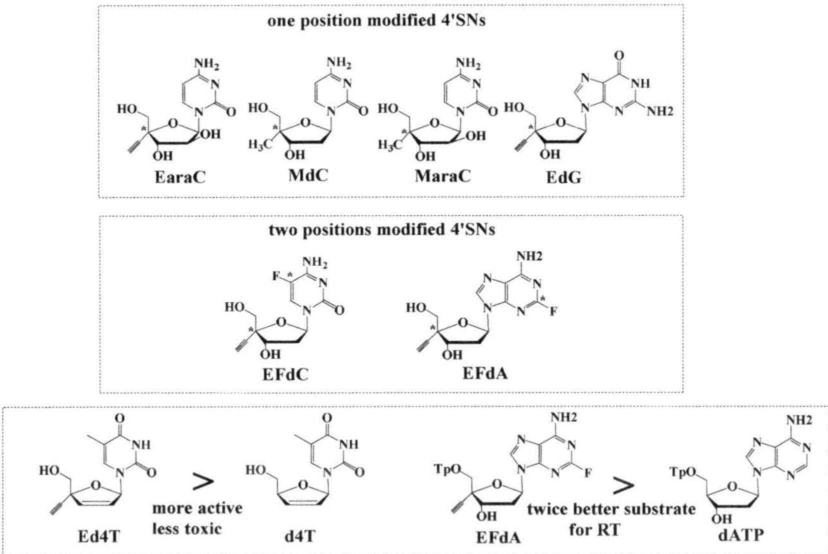

Fig. 11

The single-modification 4'SdNs depicted in Figure 11 exhibit strong anti-HIV activity but also high toxicity, suggesting that both reverse transcriptase (RT) and DNA polymerases can recognize and interact with these modified nucleosides. In contrast, the doubly modified 4'SdNs are also highly effective against HIV while demonstrating much lower toxicity. This indicates a clear preference by RT to incorporate these doubly modified 4'SdNs, whereas DNA polymerases show limited recognition, highlighting the distinct substrate selectivity between RT (an RNA-dependent DNA polymerase) and DNA polymerases (DNA-dependent DNA polymerases).

The enhanced activity of Ed4T over d4T, along with EFdATP being a more favorable substrate for RT compared to the natural substrate dATP, suggests that the 4'-C-ethynyl modification has a specific affinity for RT. This hypothesis was confirmed by an X-ray crystallography study by Yang and colleagues, which included Dr. Tanaka and Dr. Baba. The study revealed that the 4'-C-ethynyl group of Ed4T fits snugly into a hydrophobic pocket on RT, interacting with specific residues. A similar observation was made a year later with EFdA, supporting the unique interaction between the 4'-C-ethynyl group and RT.

Furthermore, EFdA was designated as a Translocation-Defective Reverse Transcriptase Inhibitor (TDRTI) because its strong affinity to RT, facilitated by both the 4'-C-ethynyl and 3'-OH groups, prevents the translocation of the 3-EFdAMP-terminated primer strand from the pre-translocation site to the post-translocation site. This blocks the addition of the next deoxy-nucleoside triphosphate (dNTP).

The reason Ed4T is less effective against HIV than EdNs may be attributed to the difficulty in phosphorylating the neopentyl-type 5'-OH in Ed4T, which lacks a 3'-OH group. This difficulty in phosphorylation and the absence of a 3'-OH group also contribute to the emergence of HIV mutants resistant to Ed4T, leading to the discontinuation of its development as a clinical drug due to resistance concerns.

"Opportunity often comes to those who are prepared!"

It was indeed a stroke of serendipity that the 4'-C-ethynyl group displayed a unique affinity for RT, and that the presence of a 3'-OH group enhanced the phosphorylation process of the 5'-OH. Such outcomes were unforeseeable.

It was through synthesis that these discoveries were made, affirming the accuracy of my theoretical frameworks. Consequently, we succeeded in developing EFdA, a compound capable of thwarting the development of resistant strains, exhibiting exceptionally potent anti-HIV activity that surpasses AZT by over 500 times and significantly exceeds the efficacy of other clinical NRTIs, all while maintaining a low toxicity profile. Therefore, EFdA holds the promise of addressing the challenges faced by existing HAART strategies.

Details on the clinical trials conducted by Merck & Co. for EFdA will be discussed subsequently. It's worth mentioning that Muftuoglu revealed an additional action mechanism of EFdA on RT, although the experimental conditions were not representative of physiological environments.

An Unexpected Development

Remarkably, on April 28, 2014, the University of Missouri announced significant news: http://research.missouri.edu/news/story.php?399
(This update was brought to my attention by one of my former students now conducting post-doctoral research at MU.)

The announcement by the University of Missouri regarding a molecule related to soy sauce flavor enhancers, later identified as EFdA, marked a significant moment. Despite initial misconceptions about its origin, the molecule's potential as a groundbreaking antiretroviral agent was clear. My subsequent communication with Professor Sarafianos clarified EFdA's synthetic nature and its design rationale, highlighting my contribution to its discovery and development. This exchange culminated in a guest lecture at the University of Missouri, where I shared the journey of EFdA's development, illustrating the collaboration and respect among scientists in advancing HIV treatment.

Initiatives for the Effective Production of EFdA

Should the FDA give the green light for EFdA's widespread application, a key challenge will be ensuring its efficient production. This is particularly relevant for Merck & Co., which has initiated clinical trials for EFdA and would benefit greatly from an efficient synthesis process. The framework for our initial synthesis of EFdA is depicted in Scheme 4.

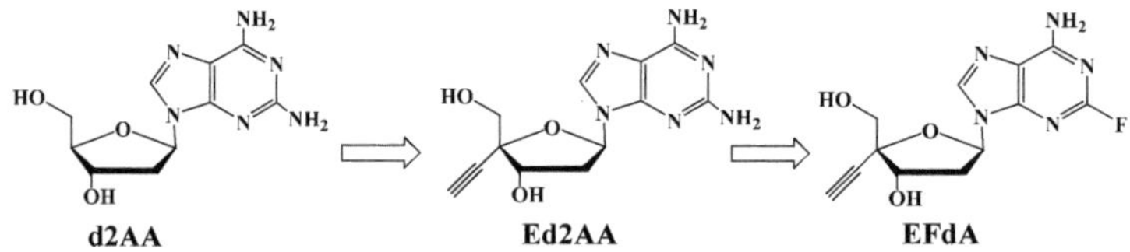

Scheme 4

The initial approach to synthesizing EFdA involved an 18-step process starting from the costly precursor 2-amino-2'-deoxyadenosine (d2AA), resulting in a modest overall yield of 2.3%. Given the necessity for a more cost-effective and efficient synthesis, we subsequently refined the process to a more streamlined 12-step sequence, achieving a significantly improved overall yield of 22%. This optimized synthesis begins with (R)-glyceraldehyde (1), as detailed in Scheme 5.

One challenge encountered during this improved synthesis was the glycosidation step, which produced a mixture of α and β-nucleosides. To obtain the desired product, it was necessary to isolate the β-isomer through column chromatography, adding a step to the purification process.

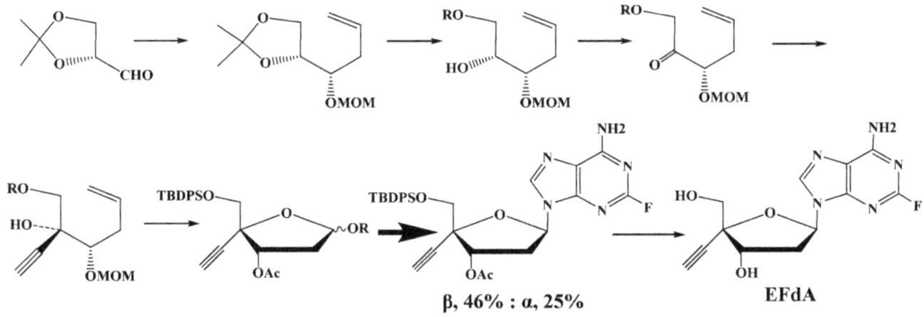

Scheme 5

Given the need for further enhancement of the synthesis process, scientists at Merck & Co. made significant improvements to our method, achieving the crystallization of the β-nucleoside directly, thus eliminating the need for chromatographic separation. Building on this advancement, they went on to develop an innovative biocatalytic cascade synthesis for EFdA, embodying the principles of green chemistry. This breakthrough was particularly remarkable because the glycosidic bond in EFdA, known for its stability to phosphorylase, had initially led me to believe that enzymatic glycosidation would be unfeasible, and thus, I had not pursued an enzymatic approach to its synthesis.

Creation of a Novel Aldol Condensation Method for Producing 4’-C-Substituted Nucleosides

The synthesis of most 4’-C-substituted nucleosides has traditionally been performed using Moffatt’s Aldol Condensation as outlined in Scheme 1. However, as previously discussed, the selective protection of the two primary alcohols (transitioning from 1 to 2 in Scheme 1) represents a pivotal step in our synthesis process. To address this, I subsequently devised a straightforward and effective aldol reaction tailored for the creation of 4’-C-substituted nucleosides, as illustrated in Scheme 6.

This new approach involves treating ketone 2, derived from diacetone-D-glucose 1, with formaldehyde and a base in tetrahydrofuran (THF), resulting in an aldol product, 3, in near-quantitative yield. The subsequent reduction of 3 with NaBH4 yields 4 quantitatively. Compound 4, featuring both a primary and a secondary alcohol, undergoes selective protection to produce compound 5 with good efficiency. The 5,6-di-O-isopropylidene group in 5 is then transformed into an aldehyde to yield compound 7, also with good efficiency. These intermediates, specifically compounds 3, 5, 6, and 7, serve as versatile precursors for synthesizing a variety of 4'-C-substituted nucleosides.

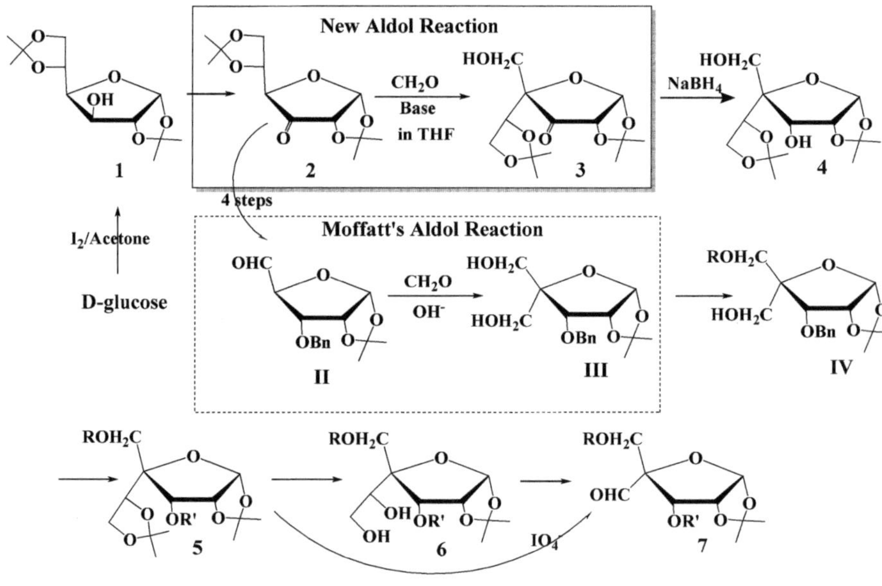

Scheme 6

In 2015, we developed a relatively efficient synthesis method for EFdA, starting from aldehyde 1. This process is detailed in Scheme 7.

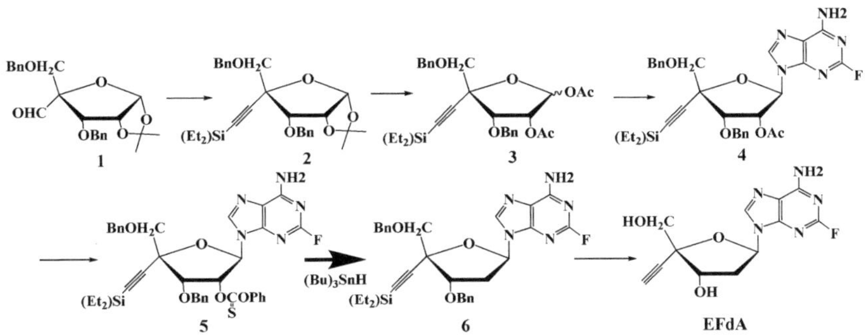

Scheme 7

The use of tributyltin hydride in the transition from compound 5 to 6, due to its high toxicity, renders the synthesis process less ideal for pharmaceutical production. Consequently, we are actively seeking alternative reagents that could facilitate the radical deoxygenation step without the associated toxicity, aiming to make the synthesis of drugs like EFdA more environmentally friendly and safer for large-scale production.

The preference for substrates by viral RNA polymerases varies distinctly from that of human RNA polymerases

A pivotal discovery from our research revealed that the substrate preferences of reverse transcriptase (RT), an RNA-dependent DNA polymerase, differ significantly from those of human DNA polymerase, a DNA-dependent DNA polymerase. This insight led to a new inquiry: Do viral RNA polymerases (RNA-dependent RNA polymerases) also exhibit distinct substrate selectivity compared to human RNA polymerases (DNA-dependent RNA polymerases)?

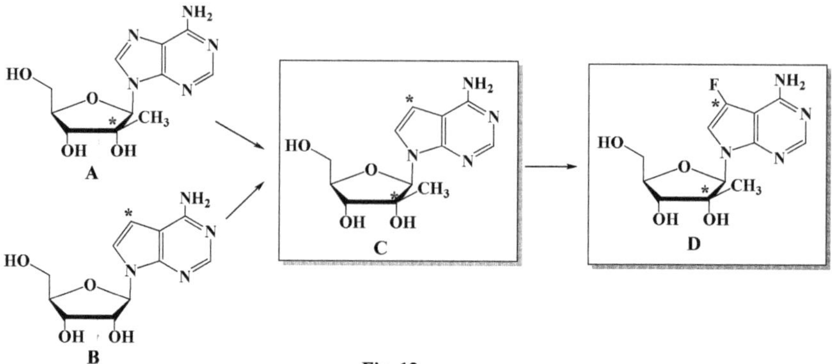

Fig. 12

To find answers, I reviewed existing literature and came across studies providing crucial insights. Eldrup and his team created a doubly modified nucleoside, 2'-C-methyl-7-deazaadenosine, combining elements of 2'-C-methyladenosine and the antibiotic tubercidine. They observed that while the singly modified nucleosides were highly effective against Hepatitis C Virus (HCV) but also highly toxic, the doubly modified nucleoside showed strong anti-HCV activity with significantly reduced toxicity. Moreover, they developed an even more potent, doubly modified anti-HCV nucleoside with lower toxicity. These findings demonstrated that further modification of toxic, singly modified nucleosides could reduce their toxicity and, importantly, that HCV RNA polymerase could process these doubly modified nucleosides whereas human RNA polymerase could not, indicating a difference in substrate selectivity.

Additionally, Smith and his team found that 4'-C-azidocytidine was active against HCV, contrasting with 4'-C-alkylcytidines which showed no activity due to their inability to be phosphorylated by kinases. They also discovered that derivatives of 4'-C-azidocytidine, including 4'-C-azidoarabinofuranosylcytidine, 4'-C-azido-2'-deoxy-2'-β-2-fluorocytidine, and 4'-C-azido-2'-deoxy-2',2-difluorocytidine, were more effective and less toxic than the original compound and acted as chain terminators for HCV RNA polymerase. Remarkably, HCV RNA polymerase was able to utilize these 2’-deoxynucleosides, highlighting an exceptional case of substrate recognition by a viral polymerase.

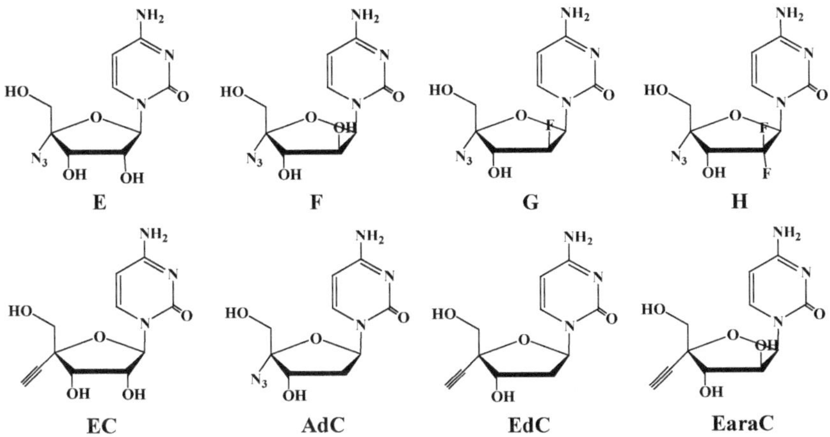

Fig. 13

Additionally, it's important to mention that both EdC and EaraC exhibit high levels of toxicity. Notably, 4’-C-azido-2’-deoxycytidine (AdC), a compound developed by Syntex, also demonstrates toxicity. Reports suggest that the development of AdC as a clinical drug was halted due to its adverse effects, leading to the eventual acquisition of Syntex by Hofmann-La Roche.

Creation of Modified Nucleoside with Anti-Influenza Activity

Upon retiring from Tohoku University at 63 in 2005, I transitioned to a lecturer position at Yokohama University of Pharmacy. Without access to a personal laboratory, I reached out to several scientists for collaborative research based on my ideas.

My interest was piqued by the observation that 4'-C-alkyl-D-ribofuranosyl nucleosides (4'SNs) lacked biological activity, presumably due to the inability of their 5'-OH group to undergo phosphorylation by cellular kinases. To test this hypothesis, 5'-O-monophosphate derivatives of 4'-C-ethynyl adenosine (EA) and 4'-C-ethynyl-2'-deoxyadenosine (EdA) were synthesized and their anti-tumor and anti-influenza activities assessed. Preliminary results of these activities are documented in Table 10, thanks to the efforts of Dr. Sugawara's team at Tokyo University of Science.

The synthesis route for the most potent anti-influenza compound, 2-(2R,3S,5R)-5-(6-amino-9H-purin-9-yl)-2-ethynyl-3-hydroxytetrahydrofuran-2-yl)-4H-benzo[d][1,2,3]dioxaphosphinine-2-oxide (EdAP), is depicted in Fig. 14. This process involved initially protecting the 5'-OH with a TBDPS group and the 3'-OH with an MPM group to obtain compound 2. This was followed by conversion to aldehyde 3 using periodate, and subsequent transformations to compound 4 and then to 6 through standard procedures. Compound 6 was then reacted with 2-chloro-4H-benzo[d][1,2,3]dioxaphosphine and diisopropylamine in acetonitrile, followed by oxidation with H2O2, yielding compound 10 with a 76% yield. The MPM group of compound 10 was removed using ceric ammonium nitrate to yield EdAP with a 75% yield.

This endeavor represents an application of the novel aldol reaction towards the synthesis of bioactive modified nucleosides. The observed biological activities of these monophosphate prodrugs, as shown in Table 10, underline the critical role of the 5'-OH phosphorylation in mediating the biological effects of 4'-C-substituted-D-ribo-nucleosides. While EA showed no biological activity, its phosphorylated counterpart, EAP, exhibited cytotoxic and anti-influenza properties, reinforcing the hypothesis regarding the importance of 5'-OH phosphorylation.

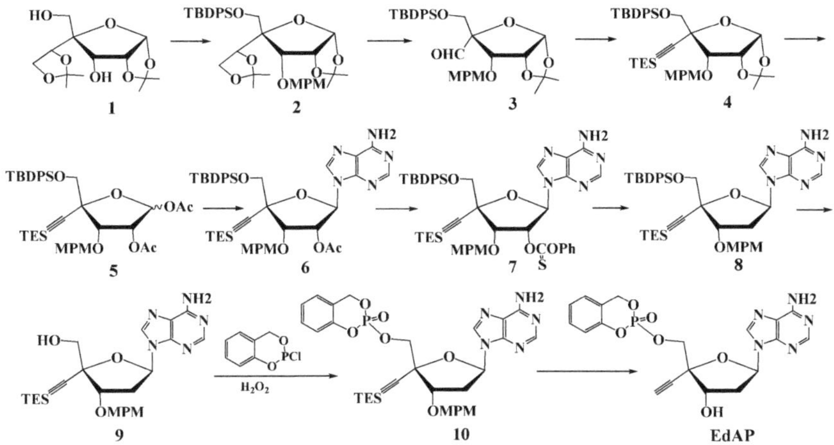

Fig. 14

Table 10. Anti-Tumor and Anti-Flu Virus Activity

Compound	Cytotoxicity (IC50 μM)			Anti-Flu Virus (H1N1) Activity
	L 1210	KB	HT-1080	
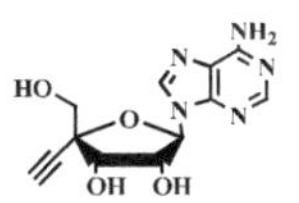 EA	>100	>100	>100	>1000
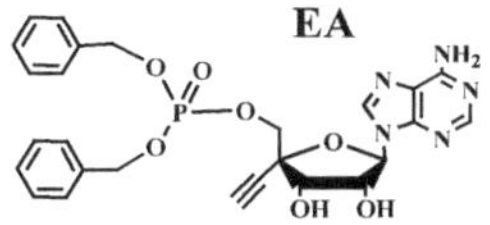 EAP	10.7	21.8	29.3	130
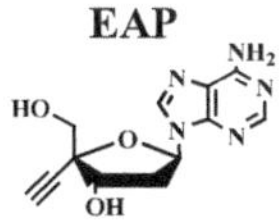 EdA	10.7	N.D	N.D	250
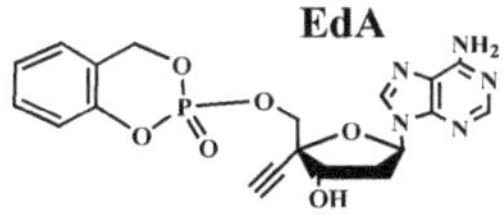 EdAP	N.D	N.D	N.D	3.8

N.D: not determined.

Given EdAP's promising in vitro anti-influenza activity, we proceeded to evaluate its in vivo efficacy using a mouse model. The initial findings are depicted in Figure 15, which illustrates that all mice subjected to influenza virus infection and not treated with any medication succumbed within 8 days. In contrast, mice that received oral administration of EdAP survived the 14-day observation period. These outcomes indicate that EdAP not only effectively combats influenza virus infection but also exhibits no toxicity to mice. Moreover, these results suggest that the influenza virus's RNA polymerase can utilize the 2'-deoxynucleoside derivative, EdAP, while the RNA polymerase in mice does not, highlighting the selective action of EdAP against the influenza virus.

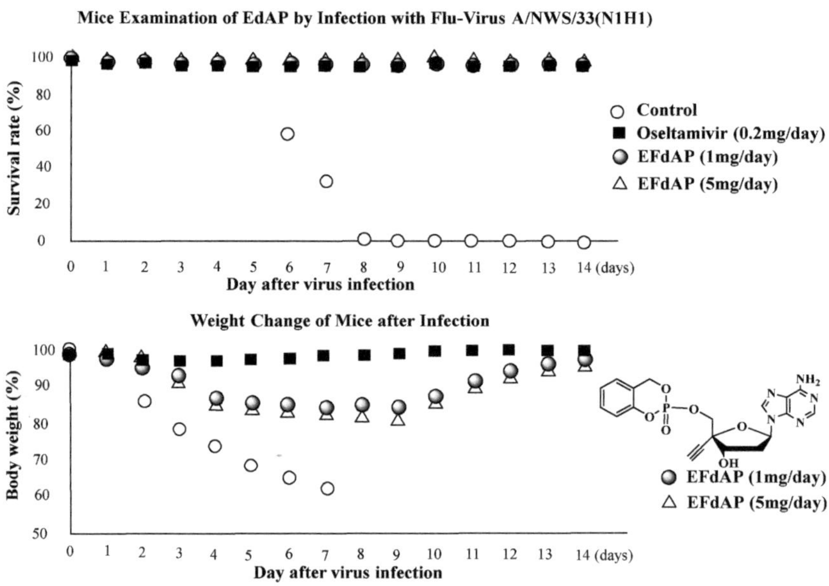

Fig. 15

These findings also highlight the differing substrate preferences between the viral RNA polymerase and the RNA polymerase in mice. Nonetheless, these results prompt several questions, including:

1. Given that the influenza virus's RNA polymerase is unlikely to utilize 2'-deoxy nucleoside derivatives, it suggests that the triphosphate form of EdA may not be the actual compound responsible for the anti-influenza activity. What then constitutes the genuine active agent in combating the influenza virus?
2. Considering EdA undergoes rapid deamination in mice, it raises the query whether EdAP also faces deamination within mice.
3. While EAP demonstrated some level of anti-influenza activity, it leads to speculation whether the triphosphate form of EA is indeed the active agent. This suggests that more effective monophosphate prodrugs might exhibit enhanced anti-influenza efficacy.

Generation of Neutralizing Antibodies 14 Days Following Infection

Figure 16 illustrates that mice treated with Oseltamivir generated relatively low levels of antibodies, suggesting that Oseltamivir effectively prevented the influenza infection. Conversely, mice treated with EdAP exhibited a significant antibody response, indicating that they were initially infected by the influenza virus but subsequently recovered due to EdAP treatment. This suggests that while mice treated with Oseltamivir might be susceptible to reinfection, those that recovered with the help of EdAP are likely to be more resistant to future influenza infections due to the higher antibody levels they developed.

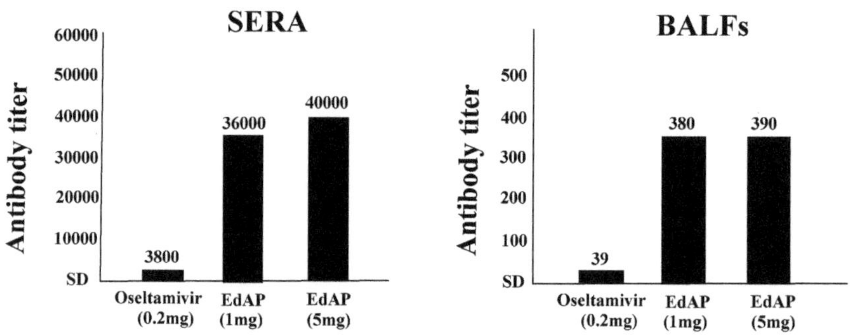

Fig. 16

Outcomes from Merck's Clinical Trials of EFdA (Islatravir)

Merck initiated clinical research on EFdA (Islatravir) in 2012 through a partnership with Yamasa Corporation. Up to September 2021, the mantra seemed to be "the more clinical trials conducted by Merck, the better the outcomes observed" with Islatravir. Highlights include:

1. Islatravir showing no significant side effects at doses up to 200mg when administered orally.
2. The compound proving effective in treating all forms of HIV.
3. Demonstrating that a daily oral dose of 0.25mg could effectively cure HIV infection.
4. A single implantation of Islatravir offering pre-exposure prophylaxis (PrEP) protection for more than a year.
5. Monthly oral doses of 60mg or 120mg for 24 weeks providing effective HIV prevention without any serious drug-related adverse events (AEs).

These findings positioned Islatravir as a potential breakthrough in HIV treatment. Unexpectedly, in September 2021, it was discovered that higher doses and the implant form of Islatravir led to reductions in CD4 cell, B-cell, and overall white blood cell counts, prompting the FDA to place a hold on HIV trials involving Islatravir. Despite this setback, Merck continued its development efforts, finding in October 2022 that the declines in CD4 and white blood cell counts were dose-dependent and reversible, and notably, the daily oral dose of 0.25mg did not trigger these declines.

Currently, Merck is exploring combination therapies involving a daily dose of Islatravir (0.25mg) with other anti-HIV medications in clinical trials.

This development suggests a dual nature of Islatravir: while high doses may pose risks akin to inducing AIDS-like conditions, it simultaneously holds promise as a leukemia treatment without the common side effect of hair loss. The mechanism behind Islatravir's impact on white cell counts is an area Merck is expected to further investigate.

Conclusion

The foundational principle behind creating antiviral drugs, along with all my hypotheses regarding the development of anti-HIV modified nucleosides, has been validated through the development of EFdA, which exhibits exceptional anti-HIV activity. This achievement underscores the distinct substrate preferences between viral RNA-dependent nucleic acid polymerases and human DNA-dependent polymerases. This difference opens up the possibility of designing antiviral modified nucleosides that target viruses specifically while being non-toxic to humans.

References

1. a) Ohrui H, (2006): *The chemical Record.* 6, 133-143,

b) (2011): *Proc. Jpn. Aca. Ser. B* 87, 53-65.

2. Mitsuya H, et al., (1985): *Proc. Natl Acad. Sci. USA,* 82, 7096-7100.

3. Murphy E. L., et al., (2001): *Ann. Int. Med.*, 135, 17-26.

4. Hung C-C., et al., (2006): *Jpn. J. Infect. Dis.*, 59, 222-228

5. Sanger F., et al., (1977): *Proc. Natl. Acad. Sci. USA*. 74, 12, 7463-5467.

6. Shuhadonlnik R.J., (1970): *Nucleoside Antibiotics, New York Wiley.*

7. Waga T, et al., (1993): *Biosci. Biotechnol. Biochem.*, 57, 1433-1438

8. Waga T, et al., (1996): *Nucleosides & Nucleotides,* 15, 287-304.

9. Youssefyeh R. D., et al., (1979): *J. Org. Chem.*, 44, 1301-1309.

10. Garegg P. C, et al., (1980): *J. Chem. Soc. Perkin Trans*, 1, 2866-2869.

11. Yamaguchi, et al., (1997): *Nucleosides & nucleotides*, 16, 1347-1350.

12. Kitano K, et al., (1997): *Tetrahedron*, 53, 13315-13322.

13. Kitano K, et al., (1990): *Bioorganic & Medicinal Chemistry Lett.*, 9, 827-830.

14. kohgo S, et al., (1999): *Biorganic & Medicinal Chemistry Lett.*, 9. 827-830.

15. Kohgo S, et al., (1999): *Nucleic acid Symposium Series*. 42. 127-128.

16. Ohrui H, et al., (2000): J. Med. Chem., 44, 4516-4625.

17.Kohgo S, et al., (2001): Biosci. Biotechnol. Biochem., 65, 1879-1882.

18. Kodama E, et al., (2001): Antimicrobial Agents & Chemotherapy.,45, 1539-1546.

19. Ohrui H, et al., (2001): *Current Drug Target-Infectious Disorder*, 1, 1-10.

20. Maag H, et al., (1992): *J. Med. Chem.,* 35. 1440-1451.

21. O-Yang C, et al., (1992): *Tetrahedron Lett.*, 33. 37-40.

22. O-Yang C, et al., (1992): *Tetrahedron Lett.*, 33. 41-43

23. Prisbe E. J, et al., (1993): *Nucleosides & Nucleotides as Antitumor & antiviral Agents. Editted by Chu C. K. & Baker D. C., New York*; Plenum Press. 101.

24. Chen M. S, et al., (1993): *Biochemistry*, 32, 6002-6010.

25. Wang G., et al., (1996): Tetrahedron Lett., 37, 6515-6518.

26. Nomura M., et al., (1999): *J. Med. Chem.*, 42, 2901-2908.

27. Beach J. W, et al., (1992): *J. Org. chem.*, 57. 2217-2219.

28. Kitano K, et al., (1999): *Bioorganic and Medicinal Chemistry*, 9. 827-830.

29. Kohgo S, et al., (2004): *Nucleosides, Nucleotides & Nucleic Acids*, 23, 671-690.

30. Hayakawa M, et al., (2004): *Antiviral Chemistry & Chemotherapy*, 15, 169-187.

31. Montgomery J. A., et al., (1969): *J. Med. Chem.*, 12, 498-504.

32. Brockman R. W, et al., (1977): *Biochemical Pharmacology*, 26, 2193-2196.

33. Hattori S, et al., (2009): *Antimicrobial Agents and Chemotherapy* 53, 3887-3893.

34. Patent: International Publication Number, 2005/090349, PCT/JP2005/005374.

35. Haraguchi K, et al., (2003): *Bioorg. Med. Chem. Lett.*, 13, 3775-3777.

36. Siddiqui M, et al., (2004): J. Med. Chem., 47, 5041-5048.

37. Yang G, et al., (2008): Antimicrob. Agents. Chemother.,52,2035-2042.

38. Michailidis E, et al., (2009): J. Biol. Chem., 18,35681-35691.

39. Nakata H, et al. (2007): Antimicrob. Agent. Chemotherapy, 51, 2701-2708.

40. Kawamoto A, et al., (2008): J. Biochem & Cell Biol., 40, 2410-2420.

41. Corb M, et al., (2012): Antimicrobial Agents and Chemotherapy, 56, 4707-4712.

42. Muftuoglu Y, et al., (2004): Antiviral Research, 106,1-4.

43. Kageyama M., et al., (2011): *Org. Lett.*, 13, 5264-5266.

44. Mclaughlin M., er al., (2017): *Org. Lett.*, 4, 926-929.

45. Yun C., et al., (2023): *J. Pharmaceutic. Biomed. Anal.*, 234, 115536.

46. Fukuyama K., (2015): *Org. Lett.*, 17, 828-831.

47. Eldrup A., et al., (2004): *J. Med. Chem.*, 47, 5284-5297.

48. Smith D. B., et al., (2009): *J. Med. Chem.*, 52, 219-223.

49. Kamata M., et al., (2020): Biosci. Biotechnol. Biochem., 84, 217-227.

www.ingramcontent.com/pod-product-compliance
Lightning Source LLC
LaVergne TN
LVHW051130210926
848482LV00001B/43